Maths
The Basic Skills

Handling Data

Veronica Thomas

Series contributors
June Haighton
Deborah Holder
Bridget Phillips

™ Nelson Thornes

D1396361

Published in 2006 by:
Nelson Thornes Ltd
Delta Place
27 Bath Road
CHELTENHAM
GL53 7TH
United Kingdom

06 07 08 09/10 9 8 7 6 5 4 3 2 1

A catalogue record for this book is available from the British Library

ISBN 0 7487 8332 6

Illustrations by Tech-Set Ltd, Gateshead, Tyne & Wear
Page make-up by Tech-Set Ltd, Gateshead, Tyne & Wear

Printed and bound in Croatia by Zrinski

Contents

Extract information

Lists

Tables

Diagrams

Pictograms

Bar charts (block graphs), numerical comparisons

Extract information skill check

Sort and classify

One criteria

Two criteria

Represent information

Handling data mock tests

Answers

Find the phone number

Here is a list of phone numbers.

Ali's Restaurant	425 851
Baz Bazaar	425 839
Brenda's Café	442 752
Chinese Cuisine	415 371
Delta Diner	424 192
Fred's Fry-up	425 173
Grub's Up	424 178

1 Write Delta Diner's phone number below.

_____

2 Write Baz Bazaar's phone number below.

_____

3 Write Chinese Cuisine's phone number below.

_____

4 Circle the word that describes how the list is ordered.

alphabetically   numerically   by date   randomly

Here is a list of phone extension numbers.

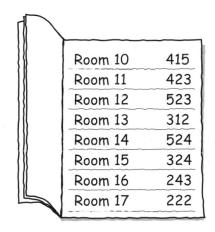

Room 10	415
Room 11	423
Room 12	523
Room 13	312
Room 14	524
Room 15	324
Room 16	243
Room 17	222

5 Write the extension number for room 16 below.

_____

6 Which room's extension number is 423?

_____

7 Which room's extension number is 524?

_____

8 Circle the word that describes how the list is ordered.

alphabetically   numerically   by date   randomly

Below is a list of the library opening times.

Opening times	
Monday	9:00 am–5:00 pm
Tuesday	9:00 am–5:00 pm
Wednesday	9:00 am–4:30 pm
Thursday	10:00 am–5:30 pm
Friday	1:00 pm–6:30 pm
Saturday	10:00 am–3:00 pm

1 When does the library open on Wednesday? _____

2 On which day is the library only open in the afternoon? _____

3 On which day is the library closed? _____

4 On which day does the library close at 6:30 pm? _____

5 What time does the library close on Saturday? _____

6 On which days is the library open from 9:00 am to 5:00 pm?

_____

7 What time does the library open on Thursday? _____

8 What time does the library close on Wednesday? _____

Extract information

Shopping list

Here is a shopping list.
Use the list to answer the questions.

pint of milk x 3
loaf of bread
lager x 4
oranges x 3
bananas x 3
apples x 4
cake x 5
tin of beans x 5

1 What does lager x 4 mean?

_____

2 What is the second item on the list?

_____

3 How many pints of milk are on the list?

_____

4 How many loaves of bread are needed?

_____

5 How many tins of beans are on the list?

_____

6 How many bananas are needed?

_____

7 Circle the word that describes how the list is ordered.

alphabetically   randomly   numerically   by date

Hospital stay

Here is a list of things to pack for hospital.

towel
shampoo
t-shirt x 3
conditioner
underwear x 3
pyjamas
toothpaste
toothbrush
soap
jeans x 2
slippers
magazine x 3
book

1 Circle the word below that describes how this list is ordered.

   alphabetically   randomly   numerically   by date

2 What is the 5th item on the list? _____

3 How many t-shirts are on the list? _____

4 What is the 6th item on the list? _____

5 What is the 4th item on the list? _____

6 How many magazines are on the list? _____

7 What is the 8th item on the list? _____

8 How many jeans are on the list? _____

Extract information

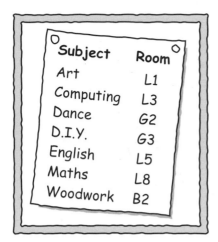

Subject	Room
Art	L1
Computing	L3
Dance	G2
D.I.Y.	G3
English	L5
Maths	L8
Woodwork	B2

1 Which room is Maths in?

_____

2 Where is Woodwork?

_____

3 Which subject is in G3?

_____

4 Where is Computing?

_____

5 Circle the word below that describes how this list is ordered.

alphabetically   randomly   numerically   by date

6 Where is Cookery?

_____

Lists can be ordered alphabetically, numerically or put in any order.

A
B
C
D
E
F
G
H
I
J
K
L
M
N
O
P
Q
R
S
T
U
V
W
X
Y
Z

Rewrite these lists in an order that you think is best.

1

Phone numbers	
Brian	483 4902
Max	839 7591
Debbie	753 5829
Imra	859 4623
Rhonda	573 3832
Fiaz	571 4525
Kath	728 6472

_____
_____
_____
_____
_____
_____
_____

2

_____
_____
_____
_____
_____
_____

Dates	
1st Jan	Meet Bev
5th Jan	Term starts
3rd Jan	Shopping
10th Jan	Dad's birthday
2nd Jan	Cinema
12th Jan	Ice skating

3

Carpet/tile services
First Floors
Aled Carpets
Cover Up
Eversure Carpets
Imagine Floors
Style and Tile

_____
_____
_____
_____
_____
_____

Tim has finished reading his book.
On the last page is a list of books by different authors.

Reggie Charles	Ancient Runes
Bill Dylan	True Dawn
Joseph Mate	Over the Hill
Joanne Moffat	In the Deep
Jen Moseley	Cakes Are In
Amos Murray	Time to Quit
May Sharella	Forest Games
Delia Tyre	Get Rich
Delia Tyre	New Release

1 Which author wrote more than
   one of these books? _____

2 How is the list ordered? _____

3 Which book did Joseph Mate write? _____

4 Who wrote Time to Quit? _____

5 Which book did May Sharella write? _____

6 Which author wrote Cakes Are In? _____

7 What is the title of Bill Dylan's book? _____

8 Which book did Joanne Moffat write? _____

9 Who wrote Ancient Runes? _____

Netball league

Below are some netball league results.
Look at the table and answer the questions.

Team	matches played	matches won	matches lost	matches drawn	goals scored
Castle Vale	6	6	0	0	38
Hilltop A	6	4	1	1	33
Rovers A	6	4	2	0	33
Cavernside	6	2	3	1	21
Hilltop B	6	2	2	2	20
Rovers B	6	2	4	0	18
West Lea	7	0	6	1	16
Beechy Head	5	1	3	1	14

1 Which team has the most goals? _____

2 Which teams have the same number of goals?

_____

3 Which team has won all their matches? _____

4 Have Rovers A drawn any matches? _____

5 Which team has scored 18 goals? _____

6 How many teams have drawn a match? _____

7 Which team has not won any matches? _____

8 Which team has played 5 matches? _____

Extract information

Dog training

Below is a register.

Subject Dog training		time 10:00–11:30 am			time 7:00–8:30 pm		
dog	owner	6/9	15/9	22/9	7/9	16/9	23/9
Choco	Eli	/	/	A	/	/	/
Jody	Eliza	/	/	/	/	/	/
George	George T	/	/	/	/	A	/
Pippa	Jess	A	/	/	/	/	/
Ben	Julie	/	A	/	/	/	A
Dipper	Mohammed	/	A	A	/	/	/
Chips	Taylor	/	/	A	A	/	A
Milo	Tim	/	/	/	/	A	/
Casper	Tom	A	/	/	/	/	/

1 Who owns Pippa? _____

2 What is the name of Tom's dog? _____

3 Who has the same first name as his dog? _____

4 Which owner was absent on 7/9? _____

5 Which owner has attended all of the lessons? _____

6 Which owner has attended least? _____

7 On what date was Eli absent? _____

8 On what date were there the most absences? _____

9 How is the register ordered? _____

Holiday in New Zealand

Average temperatures °C

	December	January	February	March	April	May
Auckland	19	20	20	19	17	14
Kerikeri	18	19	21	18	16	14
Tauranga	17	19	19	18	15	13
Wellington	15	17	18	16	14	12
Queenstown	16	15	17	15	12	7
Milford Sound	14	14	14	13	11	8
Christchurch	15	16	16	15	13	9

1 What is the average temperature in Auckland in April? _____

2 What is the average temperature in Queenstown in January? _____

3 Which place has the coldest temperature in May? _____

4 Which places have the same temperature in January and February?

_____

_____

5 When is the average temperature 16 °C in Wellington? _____

6 Which place has the hottest temperature in December? _____

7 Which place has the same average temperature in December, January and February?

_____

8 Are the temperatures in Britain the same as New Zealand in December?

_____

Extract information

Pizza Place

	8 inch	10 inch	12 inch
Margherita	£3.50	£4.10	£6.50
Milano	£4.00	£4.50	£6.80
Seafood	£4.20	£5.30	£7.50
Vegetarian	£4.20	£5.30	£7.50
Pepperoni Plus	£4.60	£5.80	£7.80
Meat Feast	£4.90	£6.00	£8.10

1   What is the cost of an 8 inch Margherita pizza? _____

2   What is the cost of a 10 inch Vegetarian pizza? _____

3   What is the cost of a 12 inch Pepperoni Plus pizza? _____

4   Which costs more: a 10 inch Seafood pizza or a 10 inch Milano?

_____

5   Which pizza costs £4.60? _____

6   Which costs more: a 12 inch Seafood pizza or a 12 inch
    Vegetarian pizza?

_____

7   Which two 8 inch pizzas cost the same?

_____

Bus times

Example

When does the 2nd bus from Cavernside get to the hospital?

	1st bus	**2nd bus**	3rd bus
Hilltop	9:00 am	1:30 pm	6:00 pm
Cavernside	9:15 am	1:45 pm	6:15 pm
Woodview	9:30 am	2:00 pm	6:30 pm
Castle Vale	9:45 am	2:15 pm	6:45 pm
West Lea	9:50 am	2:20 pm	6:50 pm
Hospital	10:00 am	**2:30 pm**	7:00 pm

Read along from Cavernside until you reach the 2nd bus column.
Read down until you reach the Hospital row.

Answer: Bus reaches the hospital at 2:30 pm

1 When is the 1st bus from Hilltop? _____

2 When does this bus get to the hospital? _____

3 When is the 3rd bus from West Lea? _____

4 When does this bus get to the hospital? _____

5 Will the 2nd bus get you to the hospital for 2:15 pm? _____

6 Will the 3rd bus get you to the hospital for 7:30 pm? _____

7 Which bus leaves Castle Vale at 9:45 am? _____

8 Which bus gets to the hospital at 10:00 am? _____

9 Which bus leaves Woodview at 6:30 pm? _____

Extract information

Banking

Below is a page from a bank statement.

Debit means pay money out.   **Credit** means pay money in.

	Details	Debit	Credit	Balance
13th Jan	Brought forward			£145.00
17th Jan	Cash withdrawal TSB	£20.00		£125.00
25th Jan	Cheque credit		£15.00	£140.00
26th Jan	4312 Credit card Smith's Birmingham	£8.23		£131.77
1st Feb	Direct debit Gas/Electric Co.	£12.40		£119.37
4th Feb	Charges	£15.00		£104.37
8th Feb	Cheque credit		£50	£154.37
8th Feb	4312 Credit card Borders	£10.45		£143.92
9th Feb	Standing order Oxfam	£8.00		£135.92
10th Feb	Cheque 3772473	£7.50		£128.42
11th Feb	Cash withdrawal TSB	£20.00		£108.42
12th Feb	4312 credit card Cinemas	£10.40		£98.02

1 How is this list ordered? _____

2 How much money was spent on 10th Feb? _____

3 Was a credit or a debit received on 1st Feb? _____

4 What was the balance on 26th Jan? _____

5 On what date was £10.45 spent? _____

6 When was the balance £135.92? _____

7 On what date did the standing order go out? _____

8 What happened on 11th Feb? _____

Extract information

Mileage chart

The chart below shows the distance in miles from one town to another. Look at the arrows to see how to use the chart.

Birmingham									
86	Bristol								
100	45	Cardiff							
281	360	365	Edinburgh						
88	162	162	206	Liverpool					
78	157	171	207	31	Manchester				
202	292	305	104	155	136	Newcastle			
61	67	100	354	155	141	257	Oxford		
76	162	178	238	69	38	133	141	Sheffield	
129	215	228	187	93	72	78	172	52	York

Example
How far is it from Edinburgh to Sheffield?
Answer: 238 miles

1 How far is it from Birmingham to Liverpool? _____

2 What is the distance from Bristol to York? _____

3 How many miles are there between Liverpool and Edinburgh? _____

4 Which towns are 38 miles apart? _____

5 What is the distance between York and Oxford? _____

6 Which towns are 67 miles apart? _____

7 Which is nearer to Liverpool: Cardiff or York? _____

8 Which two places are the same distance from Bristol? _____

9 Which place is 52 miles from York? _____

Extract information

More distances

	Aberdeen	Edinburgh	Fort William	Glasgow	Inverness	Perth	Thurso
Edinburgh	129						
Fort William	157	140					
Glasgow	146	43	116				
Inverness	105	162	67	174			
Perth	84	45	62	62	117		
Thurso	227	284	177	296	122	239	

1 What is the distance between Fort William and Perth? _____

2 Which place is 84 miles from Aberdeen? _____

3 How many miles are there between Edinburgh and Glasgow? _____

4 How many miles are there between Inverness and Fort William? _____

5 Which two places are the same distance from Perth?

_____

6 Which is nearer to Glasgow: Aberdeen or Thurso? _____

7 Which place is 162 miles from Edinburgh? _____

8 Which place is 62 miles from Glasgow? _____

9 Which is nearer to Edinburgh: Perth or Glasgow? _____

Extract information

Ice rink

This is a plan of an ice rink.

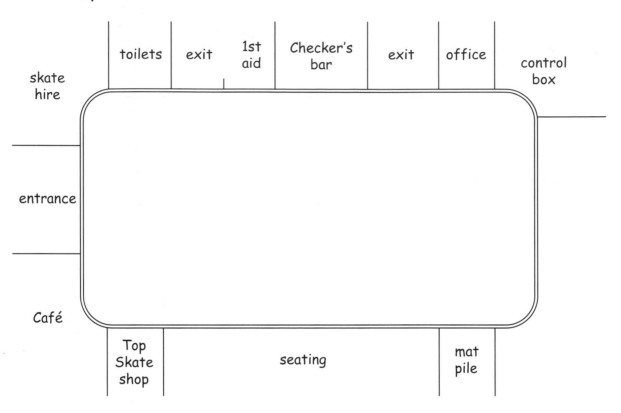

1 How many exits are there? _____

2 What is next to the mat pile? _____

3 What is between the entrance and the shop? _____

4 What is the name of the bar? _____

5 What is the name of the shop? _____

6 What is next to the control box? _____

7 What is between Checker's bar and the office? _____

8 What is between the café and the seating? _____

Extract information

Teaching room

Below is a plan of a teaching room.

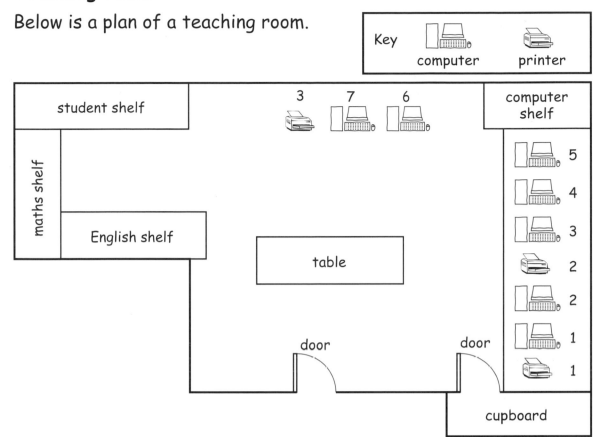

1 How many computers are in the room? _____

2 How many printers are in the room? _____

3 Which shelf is between the English shelf and the student shelf?

_____

4 What is next to printer 1? _____

5 What is between the cupboard and computer 1? _____

6 Which shelf is next to computer 5? _____

7 What is in the middle of the room? _____

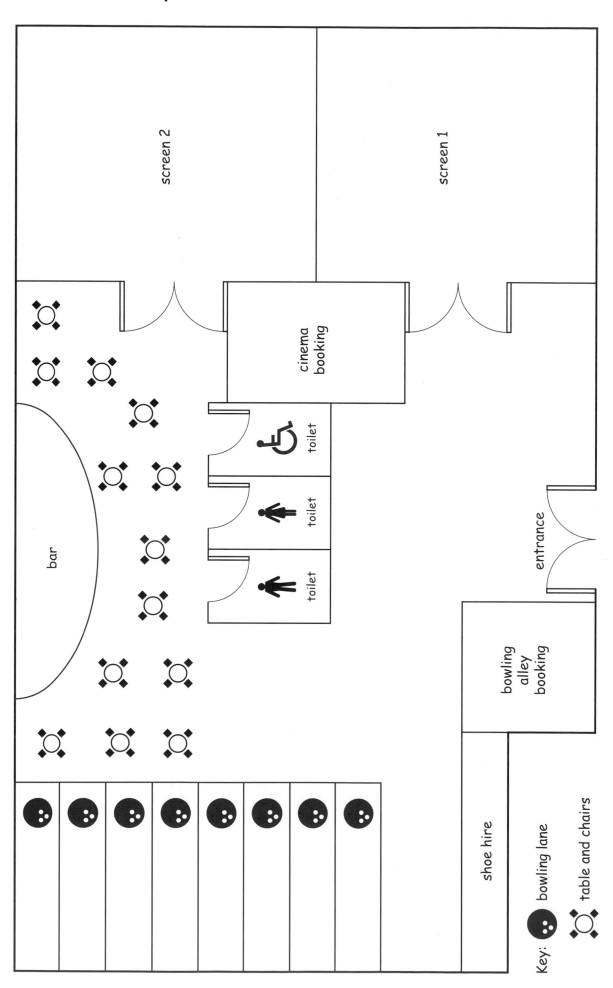

Extract information

Answer these questions using the plan of the leisure centre.

1 How many screens are there? _____

2 How many bowling lanes are there? _____

3 What is next to the shoe hire?

_____

4 How many tables are there in the bar? _____

5 You go in the main entrance and turn right. Where are you?

_____

6 How many booking areas are there? _____

7 How do you get to screen 2?

_____

_____

8 How do you get to the shoe hire?

_____

_____

Extract information

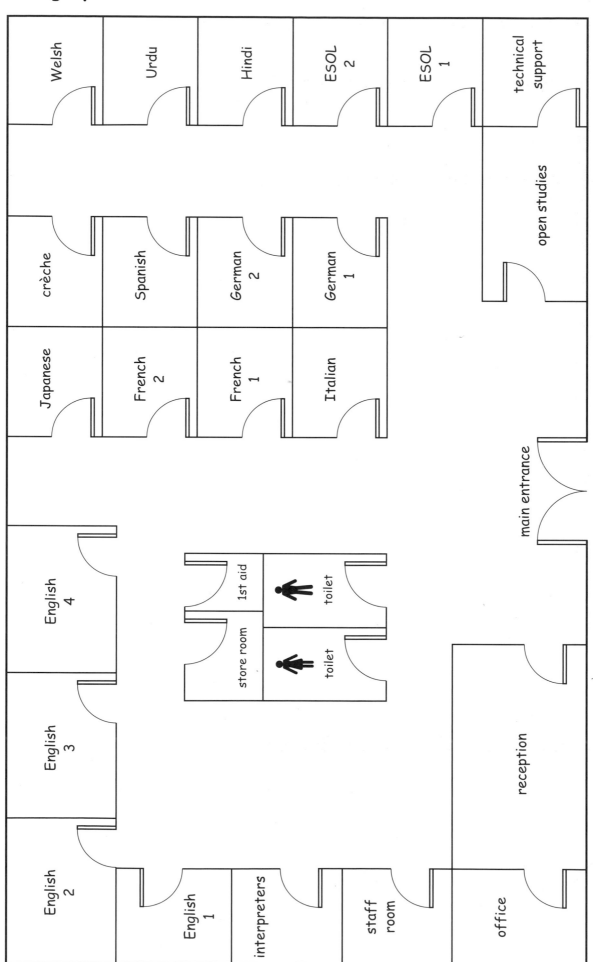

Extract information

Use the plan of the college to answer these questions.

1 What is to the left of the main entrance?

_____

2 Which rooms are the interpreters between?

_____

3 If you walk right from the entrance through the open studies room and out of the other door, where will you be?

_____

4 Come out of the Staff Room and go into the room on your left. Where are you?

_____

5 Which rooms are on either side of the Spanish room?

_____

6 Come out of ESOL 2, enter the first room on your right. Where are you?

_____

7 Which rooms are on either side of French 2?

_____

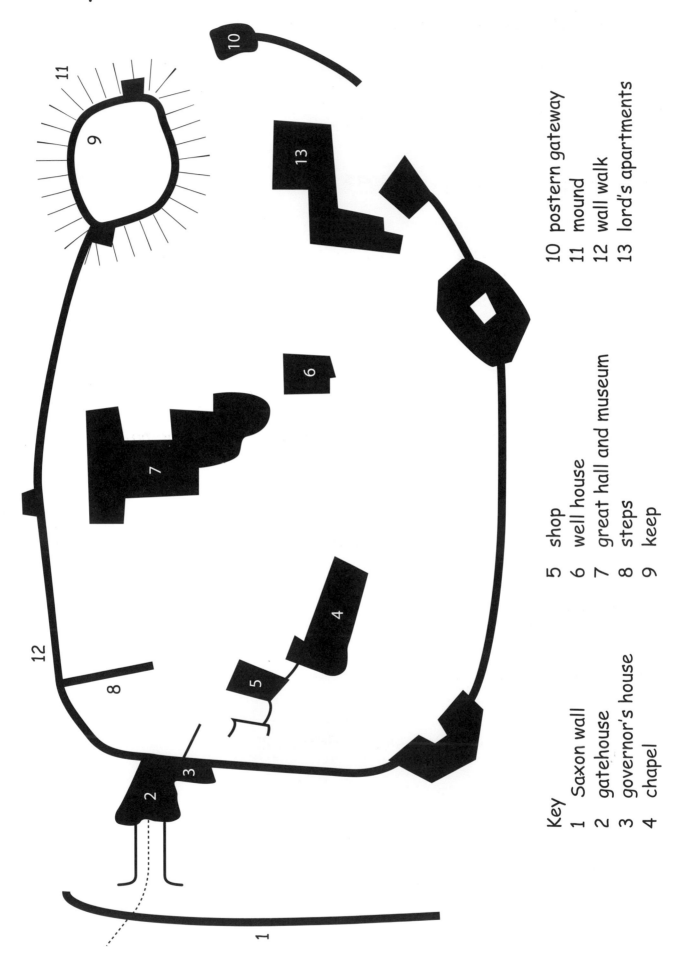

Key
1 Saxon wall
2 gatehouse
3 governor's house
4 chapel
5 shop
6 well house
7 great hall and museum
8 steps
9 keep
10 postern gateway
11 mound
12 wall walk
13 lord's apartments

Extract information

Castle plan questions E2

Use the castle plan to answer these questions.

1 What is building 2? _____

2 What number building is the chapel? _____

3 What is building number 6? _____

4 Which number building is the governor's house? _____

5 What is on the mound? _____

6 Where do the steps go to? _____

7 The great hall is in building 7. What else is in the building?

_____

8 What is to the left of all the castle buildings? _____

9 What is next to the chapel? _____

10 What is building number 10? _____

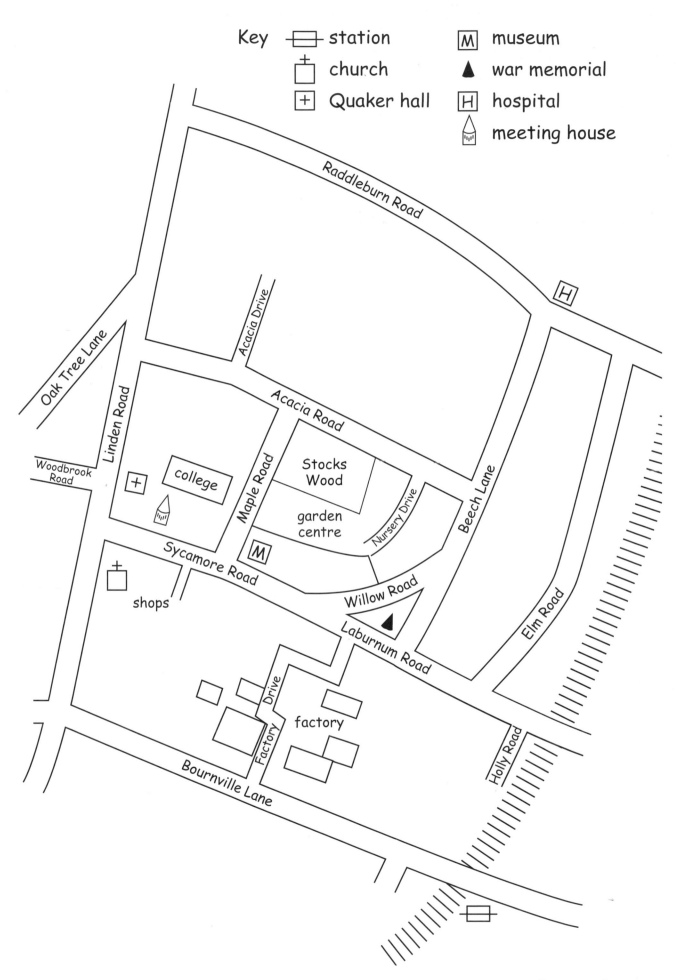

Key
- ▭ station
- ✝ church
- ⊞ Quaker hall
- Ⓜ museum
- ▲ war memorial
- Ⓗ hospital
- ⬠ meeting house

Raddleburn Road

Oak Tree Lane

Acacia Drive

Acacia Road

Linden Road

Woodbrook Road

college

Stocks Wood

Maple Road

Nursery Drive

Beech Lane

garden centre

Sycamore Road

Ⓜ

shops

Willow Road

Laburnum Road

Elm Road

Factory Drive

factory

Holly Road

Bournville Lane

Finding your way

Use the map of Bournville to answer these questions.

1 Which road is the station on? _____

2 The museum is on the corner of which roads?

_____

3 Is Elm Road near the meeting house? _____

4 Is the Quaker hall near the meeting house? _____

5 You walk from the factory to the war memorial.

Which road do you cross? _____

6 Which road does Acacia Drive join? _____

7 Stocks Wood is on which roads?

_____

8 Give directions from the station to the hospital.

_____

_____

Extract information

Below is a plan of a teaching room.
All the measurements are in metres.

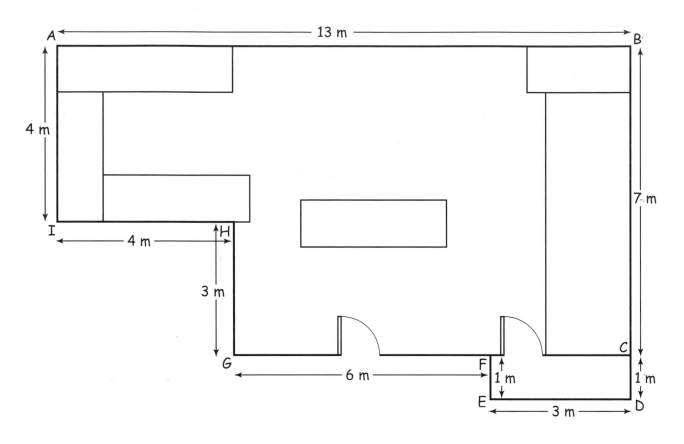

1 How long is the wall from F to G? _____

2 What is the distance between B and C? _____

3 How long is the wall between G and H? _____

4 How many metres are there between A and B? _____

5 Are there 3 metres between I and A? _____

6 How many metres are there between C and D? _____

7 What is the distance between D and E? _____

Extract information

Car colours

The pictogram shows the colour of cars in a car park.

Car colours

Key | 🚗 = 1 car

Red	🚗 🚗 🚗 🚗 🚗 🚗 🚗
Blue	🚗 🚗 🚗 🚗 🚗 🚗
Green	🚗 🚗
White	🚗 🚗 🚗 🚗
Black	🚗 🚗 🚗

1 How many cars are white? _____

2 How many cars are blue? _____

3 How many cars are green? _____

4 Which colour is there the most of? _____

5 Which colour is there more of: black or white? _____

6 Two cars are which colour? _____

Drinks

The following drinks were bought from a vending machine on Monday.

Tea   5 cups   Coffee   6 cups   Chocolate   3 cups
Soup   4 cups   Soft drink   8 cups

1 Complete the pictogram below. Remember to include a title and key.

Key [   ]   Title _____

2 Which drink sold the most? _____

3 How many soups were sold? _____

4 How many chocolates were sold? _____

5 How many soft drinks were sold? _____

6 Which drink sold the least? _____

7 5 cups of which drink were sold? _____

Extract information

TV viewing

A student asked the people in a class which TV soap they like best.
The results were:

EastEnders	8 people	Coronation Street	7 people
Neighbours	9 people	Home and Away	3 people
		Emmerdale	2 people

1 Complete the pictogram below. Remember to include a title and key.

EastEnders	
Coronation Street	
Neighbours	
Home and Away	
Emmerdale	

2 How many people like EastEnders best? _____

3 How many people like Emmerdale best? _____

4 How many people like Neighbours best? _____

5 Which was the most popular? _____

6 Which was least popular? _____

7 How many people were asked? _____

Television sales over 1 month at TV Stores

32 inch screen	
26 inch screen	
20 inch screen	
portable	
mini	

Key = 5 televisions

1 Which TV sold the most? _____

2 Which TV sold the least? _____

3 How many TVs does a symbol represent? _____

4 How many portable TVs were sold? _____

5 How many 26 inch screen TVs were sold? _____

6 How many mini TVs were sold? _____

7 How many 20 inch screen TVs were sold? _____

8 How many 32 inch screen TVs were sold? _____

Extract information

Holiday choice survey

Blackpool	
Brighton	
Bournemouth	
Scarborough	
Fishguard	

Key = 2 people

1 What does represent? _____

2 Which place is most popular? _____

3 Which place is least popular? _____

4 How many people like Scarborough? _____

5 How many people like Brighton? _____

6 How many people like Blackpool? _____

7 Which places are equally popular?

_____

8 How many people took part in the survey? _____

Cars

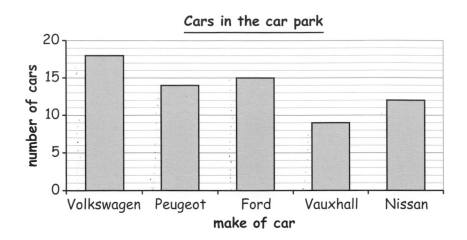

Cars in the car park

1 What is the title of this chart? _____

2 What is the most popular make of car in the car park?

_____

3 What is the least popular make of car in the car park?

_____

4 What do the numbers on the scale go up in? _____

5 What does each division on the scale represent? _____

6 Complete this table.

make of car	number of cars
Peugeot	
Volkswagen	
Vauxhall	
Ford	
Nissan	
total	

Extract information

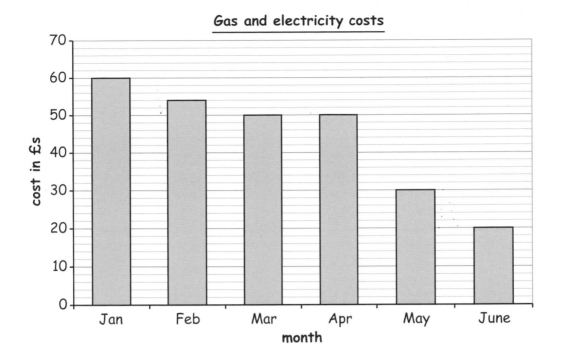

Gas and electricity costs

1 What does this chart show? _____

2 What do the numbers at the side represent? _____

3 What does the scale go up in? _____

4 Which months cost the same? _____

5 Which month was the cheapest? _____

6 Which month was the most expensive? _____

7 How much more expensive was January than February? _____

8 How much cheaper was June than January? _____

9 How much cheaper was June than May? _____

Café drinks

The chart shows the amount of drinks sold in the college café at break.

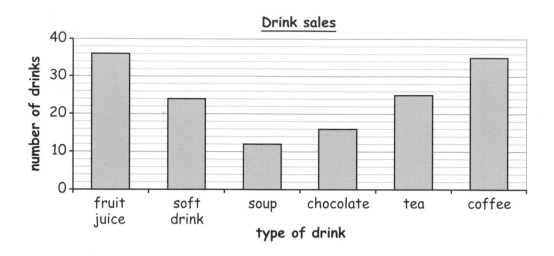

1 What is the title of this chart? _____

2 What does each division on the scale represent? _____

3 Which drink was most popular? _____

4 Which drink was least popular? _____

5 In the table below list the drinks in order of how many were sold. Start with the most popular.

6 Complete the table.

type of drink	number of drinks sold

Extract information

College courses

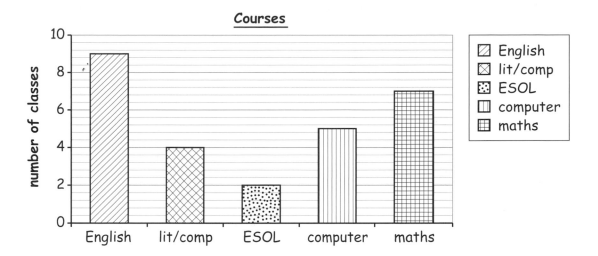

1 What is this bar chart about? _____

2 How many lit/comp classes are there? _____

3 How many computer classes are there? _____

4 Which subject has the most classes? _____

5 How do you know this? _____

6 How many maths classes are there? _____

7 How many ESOL classes are there? _____

8 How many classes are there altogether? _____

Extract information

Smoking survey

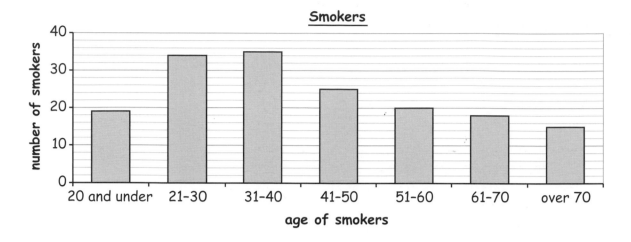

1 Complete this table.

age of smokers	number of people
20 and under	
21–30	
31–40	
41–50	
51–60	
61–70	
over 70	

2 How many people smoke who are under 31 years old? _____

3 How many people smoke who are over 60? _____

4 How many people smoke who are between 41 and 60? _____

5 Which age group has the most smokers? _____

6 How many people took part in the survey? _____

Extract information

Fruit

A student asked people what their favourite fruit was.
The results were:

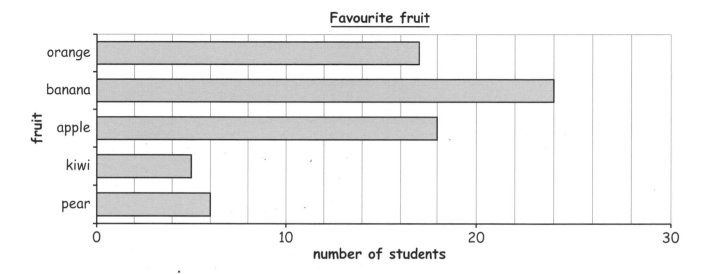

Favourite fruit

1 What does the chart show? _____

2 Which fruit is most popular? _____

3 Which fruit was chosen by 6 people? _____

4 What do the numbers on the scale go up in? _____

5 What do the grid lines go up in? _____

6 How many more people chose apples than oranges? _____

7 How many people took part in the survey? _____

Films

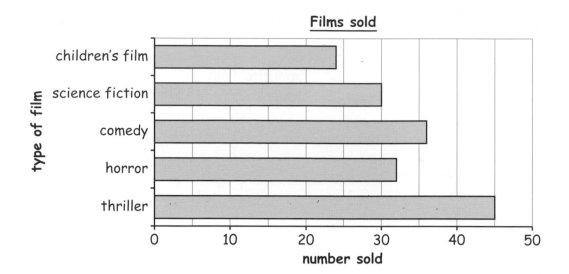

1 What is the bar chart about? _____

2 What type of film sold most? _____

3 What does the scale go up in? _____

4 Which type of film sold 36? _____

5 How many horror films were sold? _____

6 Which type of film sold the least? _____

7 Why do you think that was? _____

   _____

Extract information

Smokers

This chart compares the number of men and women who smoke.

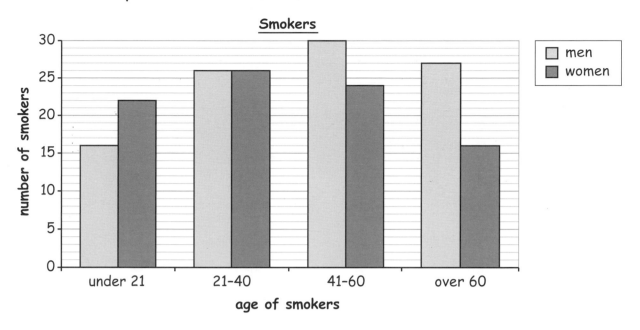

1 Between which ages do the same number of men and women smoke?

_____

2 At what age do more women smoke than men? _____

3 Which age group has the most men? _____

4 Which age group has the smallest number of men? _____

5 Complete this table.

smokers	number of men	number of women	total
under 21			
21–40			
41–60			
over 60			
total			

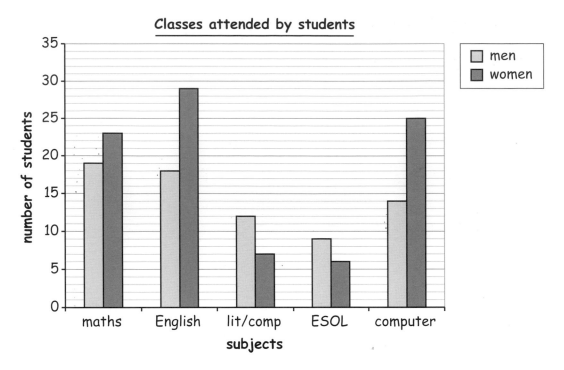

1 What does the chart show? _____

2 Why are the bars in two different colours? _____

3 Who attends more maths classes: men or women? _____

4 Which subject is most popular with women? _____

5 Which subject is most popular with men? _____

6 Do more women attend ESOL classes than men? _____

7 How many women attend classes altogether? _____

8 How many men attend classes altogether? _____

9 Which subject is least popular? _____

What is missing?

3 of the following charts have something missing.

Write in the missing things.

1

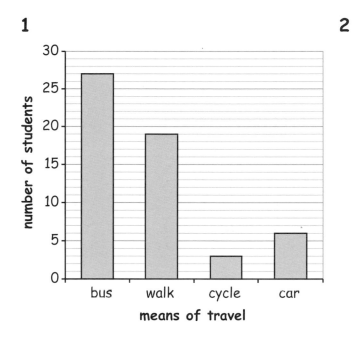

2

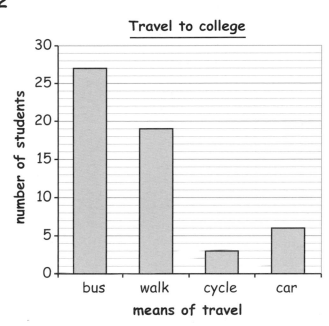

3

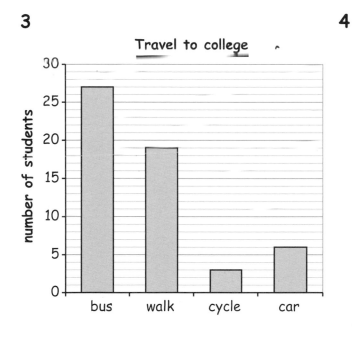

4

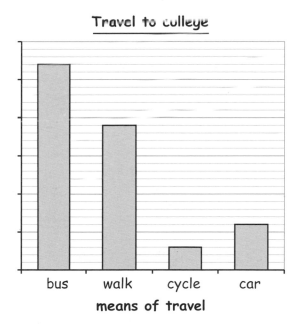

Extract information

Extract information skill check 1

A plumber is needed

p1-6

1 What is the number for Best Fit?

_____

Plumbers	
All Jobs	472 8188
Best Fit	473 3452
Fix It Pipes	471 6217
Hilltop Plumbers	472 3369
Mass and Son	471 6726
Watertight	472 9899

2 Which plumber has the number 471 6726?

_____

3 Circle the word that describes how this list is ordered.

alphabetically   numerically   random

Here are some instructions to get to 38 Bournville Lane.

```
1  Get on the M42
2  At junction 2 join the A441
3  Head to King's Norton
4  At King's Norton head to Cotteridge
5  At Cotteridge shops head to Bournville
6  Take the 2nd right into Bournville Lane
7  Number 38 is on the right
```

4 What is the 2nd instruction?

_____

5 Which instruction is _Head to King's Norton?_

_____

6 Circle the word that describes how this list is ordered.

alphabetically   numerically   random

Look at the plumber's bill.

7 What is the labour cost?

_____

8 What costs £25.00?

_____

New pipe	£8.00
Tape	£1.50
Call out	£25.00
Labour	£10.00
Vat	£7.79
Total	£52.29

Extract information skill check 2

Jackie is taking her 6 year old son to the zoo on Saturday.

Zoo Opening Times

	Open	Close
Monday	10 am	4 pm
Tuesday	10 am	4 pm
Wednesday	9 am	12 pm
Thursday	10 am	4 pm
Friday	10 am	4 pm
Saturday	10 am	5 pm
Sunday	11 am	5 pm

 p7

1 What time will the zoo open?

_____

2 When will the zoo close on Saturday?

_____

Bus 45

Hogshead	9:15	9:35	9:55
Sandwell	9:25	9:45	10:05
Rockvale	9:30	9:50	10:10
High Street	9:40	10:00	10:20
West Gate	9:45	10:05	10:25
Zoo	10:00	10:20	10:40

p8-13

3 Jackie gets the 1st bus from Sandwell. What time is this?

_____

4 What time will they get to the zoo?

_____

5 What is the cost of Jackie's ticket?

_____

Zoo Tickets

Adults	£9.00
Children 5-16	£6.00
Children under 5	free
Family ticket	£20.00

6 What is the cost of her son's ticket?

_____

7 Is it cheaper for Jackie to get a family ticket?

_____

This is a plan of one reptile house

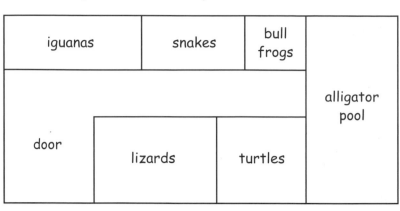

p16-26

8 Which animals have the largest area?

_____

9 Which animals are next to the door?

_____

10 Which animals are between the iguanas and the bull frogs?

_____

This chart shows the number of tickets sold.

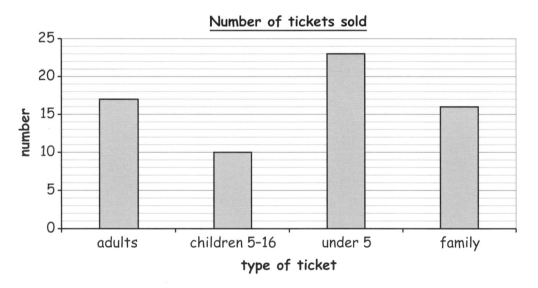

p32-41

11 How many family tickets were sold? _____

12 Which type of ticket sold the most?_____

13 Which sold more: adult tickets or family tickets? _____

Extract information

14 What is missing from this chart? Circle one of the following:

scale   title   labels

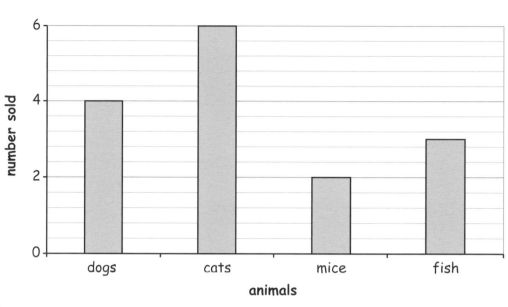

p27-31

This pictogram shows the number of bowls of fruit eaten by animals in the zoo.

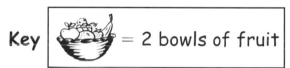

Key [bowl image] = 2 bowls of fruit

monkeys	[bowl] [bowl]
elephants	[bowl] [bowl] [bowl] [bowl]
chimps	[bowl] [bowl] [bowl]
giraffes	[bowl]

15 What does mean? _____

16 How many bowls of fruit did the elephants have? _____

17 Which animals had 4 bowls of fruit? _____

Extract information

Extract information skill check 1 answers

1 473 3452

2 Mass and Son

3 alphabetically

4 At junction 2 join the A441

5 instruction 3

6 numerically

7 £10.00

8 Call out

Extract information skill check 2 answers

1 10 am

2 5 pm

3 9:25

4 10:00

5 £9.00

6 £6.00

7 no

8 alligators

9 lizards

10 snakes

11 16

12 under 5

13 adults

14 title

15 2 bowls of fruit

16 8 bowls

17 monkeys

Food and drink

Below is a list of food and drink items.

Write them in 2 lists - a list of food and a list of drinks

lager
bread
coke
coffee
carrots
crisps
milk
fish
water
curry
lemonade
pizza
water
spaghetti

food	drinks

Clothes go in the Oxfam recycling bank.

Paper goes in the paper bank.

Draw an arrow from each object to the correct recycling bank.

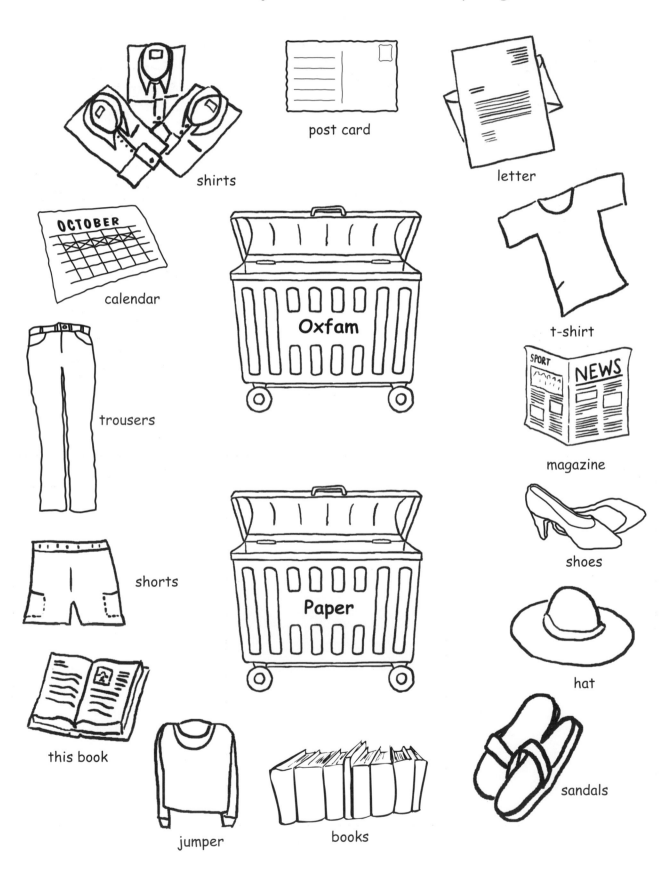

shirts

post card

letter

calendar

Oxfam

t-shirt

trousers

magazine

shorts

Paper

shoes

this book

hat

jumper

books

sandals

Food shopping

You go shopping. The things you need are shown below.

You go to the butcher, baker and greengrocer (vegetable shop).

Make a list of what you buy from each shop.

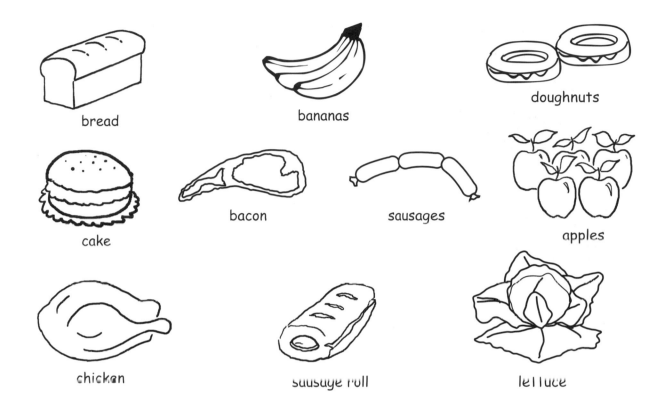

bread

bananas

doughnuts

cake

bacon

sausages

apples

chicken

sausage roll

lettuce

baker	butcher	greengrocer

Tools

Garden tools go in the garden shed.

DIY tools go in the garage.

Make a list of where the tools go.

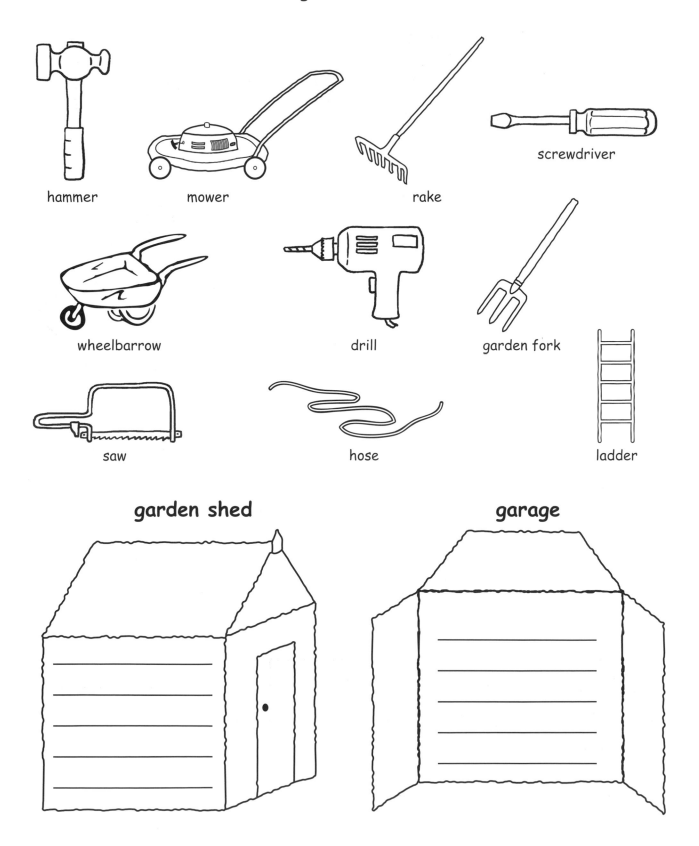

hammer

mower

rake

screwdriver

wheelbarrow

drill

garden fork

saw

hose

ladder

garden shed

garage

Sort and classify

Shapes

Some of the shapes below are 2-D (flat).
Others are 3-D (not flat).

Colour the 2-D shapes in.

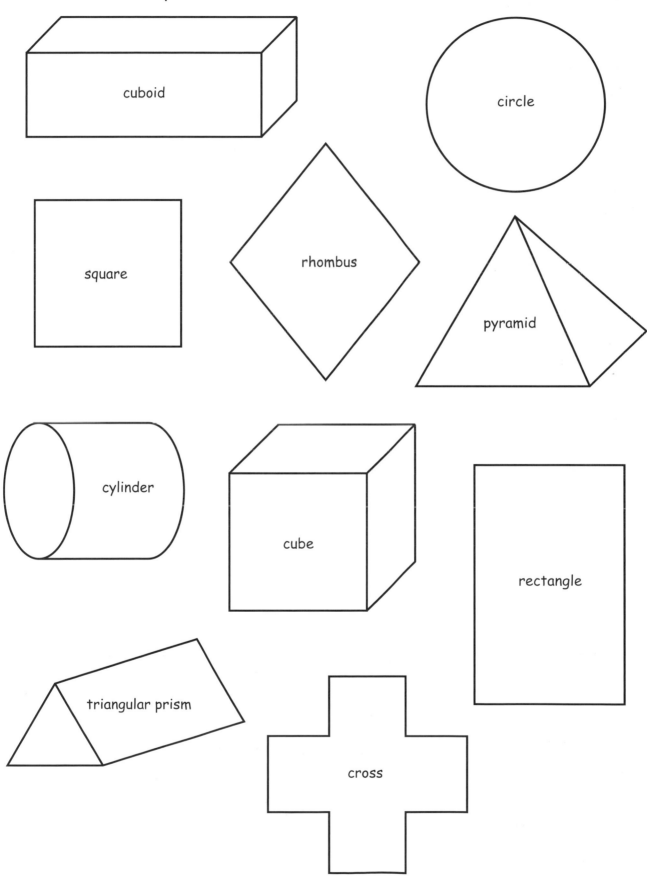

Vehicles

1 Colour the 2-wheel vehicles red.

2 Colour the 3-wheel vehicles blue.

3 What type of vehicles are not coloured?

motorbike

sports car

penny farthing

tricycle

saloon car

3-wheeled car

bicycle

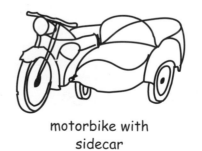

motorbike with
sidecar

taxi

Sort and classify

1 College courses have been sorted into three groups.

Education, sport or art/craft.

How have they been sorted?

By: day   subject   time   ⟵——— Circle one of these

2 How have these cans been sorted?

By: size   colour   contents   ⟵——— Circle one of these

3 How have these clothes been sorted?

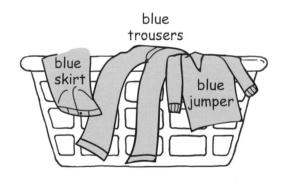

By: size   colour   age   ⟵——— Circle one of these

The clocks all tell different times.

Circle the clocks that are digital **and** on the hour.

1

2

3

4

5

6

7

8

9

10
08:15

11
09:00

12

13
04:00

14

15

16
03:30

Sort and classify

Jumbled shapes

Below are lots of different shapes.

Colour shapes with curved edges and straight edges blue.

Colour shapes with straight edges and less than 5 sides red.

Colour shapes with straight edges and more than 5 sides green.

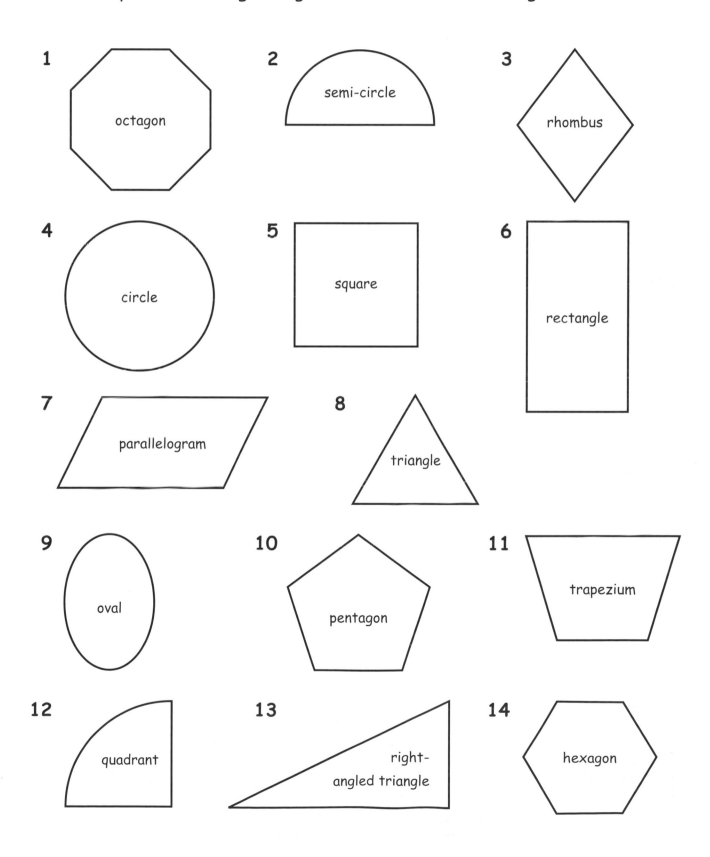

1 octagon

2 semi-circle

3 rhombus

4 circle

5 square

6 rectangle

7 parallelogram

8 triangle

9 oval

10 pentagon

11 trapezium

12 quadrant

13 right-angled triangle

14 hexagon

Sort and classify

A café records what people eat.

1 How many people have just pizza and chips? _____

2 How many meals included both pizza and beans? _____

3 How many people have just salad and chips? _____

Sort and classify

Clothes

colour

	red	blue	green	white	black
jumper	2	3	1	0	1
trousers	1	3	0	1	3
skirt	1	3	1	0	2
dress	1	2	1	1	2
t-shirt	2	3	0	2	1
hat	0	0	1	0	1

clothes

The table shows the clothes in a wardrobe.

1 How many blue trousers are there? _____

2 Are there any green t-shirts? _____

3 Which colour skirt is there more of: black or blue? _____

4 What colours of hats are there? _____

5 Which colour t-shirt is there most of? _____

6 How many white trousers are there? _____

7 How many black jumpers are there? _____

8 Is it true that there are 2 red jumpers? _____

9 Are there any white dresses? _____

classes

	agility	dog handling	heel work	obedience	fly ball
collie	15	5	12	14	16
poodle	1	4	1	4	3
bulldog	6	7	9	7	10
spaniel	4	3	4	6	8
labrador	0	5	6	10	6
terrier	0	3	3	6	4

dogs

1 How many collies are there in the fly ball class? _____

2 In the obedience class how many labradors are there? _____

3 How many terriers are there in the heel work class? _____

4 How many spaniels are there in the dog handling class? _____

5 Which class doesn't have any labradors? _____

6 How many bulldogs are in the agility class? _____

7 How many collies are there in total? _____

8 Which classes have four spaniels in them? _____

Favourite drinks

Ask some students what their favourite drink is.

Record the results in the tally chart.

Remember	|||| = 4
	ⅣⅢ = 5

drink	tally	total
beer		
coffee		
tea		
coke		
fruit juice		
chocolate		
other		

1 What is the most popular drink? _____

2 Which is the least popular? _____

3 Do more students like tea or coffee? _____

4 Do more people like beer or chocolate? _____

5 How many students took part in the survey? _____

Fast food

Ask some people what their favourite fast food is.

Record your results in the tally chart.

Fill in the total column.

fast food	tally	total
McDonalds		
Burger King		
Subway		
fish and chips		
pizza		

1 Do more people like McDonalds or Burger King best? _____

2 How many people like Subway best? _____

3 How many people like pizza and fish and chips best? _____

4 Which is the most popular fast food? _____

5 Which is the least popular fast food? _____

6 How many people did you ask altogether? _____

7 Should there have been an 'other' option? _____

8 Do you think the results are reliable? _____

9 Why? _____

Sort and classify

Holiday park

Here is a tally chart for you to fill in.

A survey asked people which holiday park they like best.

Their answers are listed below.

Coast	Bay	Coast	Bay	Hill top	Town
Hill top	Town	other	Coast	Hill top	Bay
Bay	Town	Bay	other	Bay	Hill top
Coast	Hill top	other	Coast	Town	Hill top
Bay	Town	other	Town	Bay	Hill top
Town	Hill top	Bay	other	Coast	other
Bay	Town	Hill top	Town	other	Hill top
Bay	Coast	Town	Hill top	Bay	other
Town	Bay	Hill top	Bay		

The first 3 answers have been put in the tally chart below.

It is a good idea to cross out the answers as you use them.

Complete the tally chart.

holiday park	tally	total
Coast	||	
Bay	|	
Hill top		
Town		
other		

Music

A survey asked people to name their favourite type of music.
Here are the answers:

classical	pop	classical	pop
other	jazz	rock	jazz
rock	other	classical	jazz
pop	pop	rock	rock
pop	other	pop	jazz
classical	jazz	other	classical
rock	other	pop	jazz
pop	rock	jazz	rock
pop	jazz	rock	pop
jazz	pop	other	classical
other	pop	rock	jazz
rock	other	jazz	pop
classical	pop	rock	jazz

Complete the tally chart below.

Remember to cross out the answers as you use them.

music	tally	total
classical		
rock		
jazz		
pop		
other		

Sort and classify

Questionnaire

Look at this library questionnaire.

Tick the boxes.

How old are you?   Under 16 ☐

under 25 ☐

under 50 ☐

under 80 ☐

How often do you borrow items from the library?

Once a week   ☐

More often   ☐

At what time of day do you use the library? _____

Do you borrow books? ☐

Do you borrow CDs? ☐

Do you use the computer at the library? ☐

What type of book do you like best? ☐

How could you improve this questionnaire? _____

_____

_____

_____

_____

Pets

Suppose you want to know which pets people have.

1 Circle the question you think is best.

Do you like dogs?   yes ☐   no ☐

Do you have a cat?   yes ☐   no ☐

What pets do you have? _____

Do you have a pet?   yes ☐   no ☐

2 Circle the way you will ask questions.   verbally

                                                    questionnaire

3 Circle the type of people you will ask.   students

                                                    adults

                                                    children

4 Circle the best time to ask questions.

during the day   late at night   early in the morning

5 Circle the best places to ask questions.

at a supermarket   at an animal rescue centre

at a pet shop   on the train

at work

Sort and classify

Sort and classify skill check 1

p47-53

1 Write these names in two lists:
men's names and women's names.

men's names	women's names

Ben   Sue
James   Jill
Jenny   Tom
Mike   Jane

2 Draw a circle around the baby items.

3 Tick the shapes that have straight sides.

   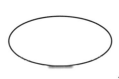

4 Tick the clothes to go in the white wash.

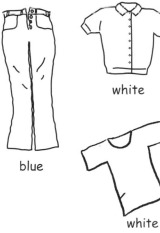  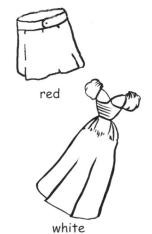 

white   red   blue

blue

white   white

white wash

5 Tick the items you can drink.

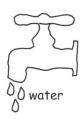  

water   paint   coke   beans   dog food

Skill check

Sort and classify skill check 2

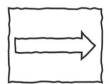 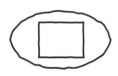

p54-58

1 Put a tick beside the big dogs with no tails.

2 Put a cross beside the small dogs with tails.

3 Put a tick beside the shapes with straight sides and arrows.

4 Put a cross beside the squares with circles inside.

5 A person stores things for the house in the garage and things for the garden in a garden store.
The garden store is 1 metre high.

Tick the items that will go in the garden store.

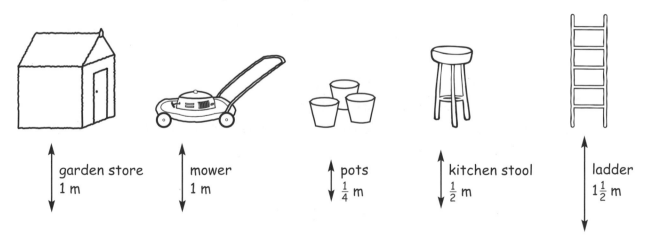

	blue	red	white	black
Fiat	2	3	1	3
Ford	4	1	3	2
Nissan	1	2	0	1
Vauxhall	2	2	1	0

6 How many red Fiats are there? _____

7 How many blue Fords are there? _____

p59-64

8 A survey asked people which activity they liked best.
Here are the results.

~~TV~~	~~bowling~~	other	other	gym	TV	art/craft
art/craft	TV	gym	other	art/craft	other	bowling
TV	other	gym	gym	bowling	TV	other
TV	other	art/craft	TV	other	art/craft	bowling

Complete the tally chart.
The first two have been
done for you.

activity	tally	total
art/craft		
bowling			
gym		
TV			
other		

9 You want to know which day students prefer to attend college.
Tick the question you would use.
  Which day do you attend college?
  Do you attend college every day?
  Is there one day which is better for you to attend than others?
  Which day would you like to attend classes on?

10 How would you improve this question?

  Are you   under 18 ☐ _____
        18–24 ☐ _____
        24–30 ☐ _____

Sort and classify skill check 1 answers

1 men's names = James, Ben, Mike, Tom
women's names = Sue, Jill, Jenny, Jane

2

3

4

5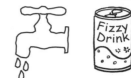

Sort and classify skill check 2 answers

1

2

3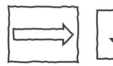

4

5 mower, pots

6 3

7 4

8 art/craft ⊮⊮ (5)   bowling |||| (4)   gym |||| (4)   TV ⊮⊮ || (7)   other ⊮⊮ ||| (8)

9 Which day would you like to attend classes on?

10 answer should include:
change either 18–24 or 24–30 so that 24 does not appear twice
include information about what goes in the boxes

Writing a list

If you want more than one item use a times sign (x) and then the number you want.

Example   'Tin of dog food x 3' means 3 tins of dog food.

Write a shopping list for these items.

6 apples

2 doughnuts

1 lettuce

3 cartons of milk

5 cans of beer

4 bananas

Shopping list

_____

_____

_____

_____

_____

_____

1 Order the lists of names and numbers below so that your friend can find the phone numbers easily.

Bill 642814

Mohammed 346282

Soumia 382848

Rajaa 079774435534

Paul 673828

Leila 672453

Vicky 678929

David 079563728463

2 You want your friend to look at some pages in a book.
List the pages in order.

p. 13   p. 18   p. 8   p. 4   p. 7   p. 21   p. 16   p. 11

3 Now list this column of dates in order:

5 March

2 April

9 April

26 April

1 March

12 March

24 April

17 April

21 March

29 March

Represent information

Dog kennel jobs

Ben and Jill work at a kennels.

They have drawn up a rota of the jobs.

Monday	Tuesday	Wednesday	Thursday	Friday	Saturday	Sunday
walk dogs 1	walk dogs 1	walk dogs 1	walk dogs 1	walk dogs 1	walk dogs 1	walk dogs 1
clean runs	clean runs	clean runs	clean runs	clean runs	clean runs	
see vet	clean visitor room	see vet	clean visitor room	see vet	clean visitor room	
feed dogs	feed dogs	feed dogs	feed dogs	feed dogs	feed dogs	feed dogs
walk dogs 2	walk dogs 2	walk dogs 2	walk dogs 2	walk dogs 2	walk dogs 2	walk dogs 2

1 On which day do they only feed and walk the dogs? _____

2 How many times in a week do they clean the visitor room? _____

3 How many times do they see the vet in a week? _____

4 How many times do the dogs get walked in a week? _____

5 Choose a colour for each person.

Key	
Ben	☐
Jill	☐

Colour code the rota so that the jobs are split as fairly as possible between them.

Office spending

Below is a list of the office spending over two weeks.

envelopes	£2.40	£1.50
stamps 1st class	£3.00	£1.50
stamps 2nd class	£2.10	£2.10
paper	£2.40	£3.00
pens	£3.00	£2.00
pencils	£1.50	_____
milk	£1.42	£1.42
coffee	_____	£5.95
sugar	£0.68	£0.68

Put this information in a table.

Give the total spending for each item.

Add a title and headings.

Represent information

Timetable

The times for the vet to see the dogs are listed below.

Monday	Spot 9-11	Bruno 1-2	Prince 2-3
Tuesday	off all day		
Wednesday	Lady 9-10	Sandy 11-1	Bob 1-2
Thursday	off all day		
Friday	Ben 10-12	Trixie 1-3	

Put this information into the timetable below.

	Monday	Tuesday	Wednesday	Thursday	Friday
9-10					
10-11					
11-12					
12-1					
1-2					
2-3					

Table top sale

Tables are set around a hall.

The door and table A are shown.

Read the information below and label the other tables.

B is opposite the door.

C is on the left of the door.

D is opposite C.

E is opposite A.

F is next to A.

G is opposite F.

H is to the right of the door.

I is opposite H.

A

door

Design a kitchen

Below is a plan of a kitchen.

Draw these on the grid:

    A sink in the shape of a rectangle.

    An oven in the shape of a square.

    A washing machine in the shape of a square.

    A worktop in the shape of a rectangle.

    A table in the shape of a circle.

Either label the items or use a key to show what they are.

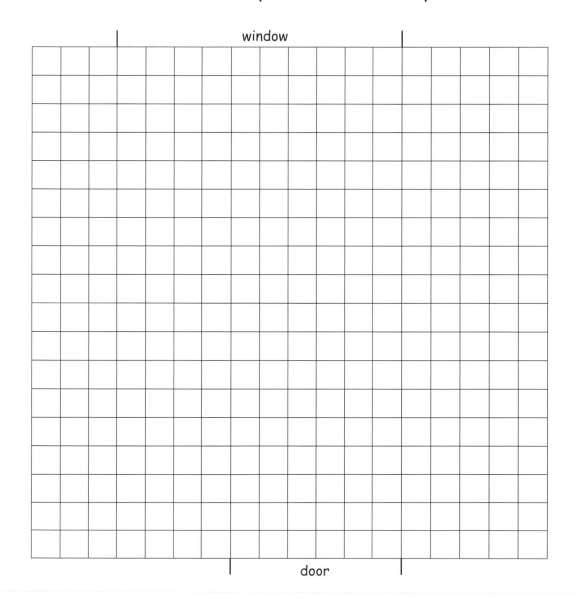

Represent information

The maths books are colour-coded by topic.

Number books have a blue sticker.

Measuring books have an orange sticker.

Handling data books have a green sticker.

Mark the books below in the correct colour.

1 Draw a plan of your bedroom.

Draw the shape of your room.
Mark the windows and door.
Show the bed and other items.

You could use **shapes** and **label** them, or use a **key**.

Example: (**bed**)   <u>or</u>   **Key** (   ) = bed

2 Write down 3 questions about your bedroom to ask somebody who has never been to your house.

_____

_____

_____

Could they answer these questions using your plan? _____

3 Draw a plan of your bathroom.

Draw the shape of the room.
Show the door and window.
Show the toilet, sink, bath or shower.

4 Write down 3 questions about your bathroom to ask somebody who has never been to your house.

_____

_____

_____

Could they answer these questions using your plan? _____

The tally chart gives information about classes.

classes	tally	total
maths	⊞ l	
English	⊞ lll	
computing	⊞ ll	
ESOL	ll	
typing	⊞	

1 Fill in the totals.

2 Draw a pictogram to show the data.

Use △ for a class.

> **Remember the title and key.**

Below is another tally chart.

leisure classes	tally	total
art	⊞ ll	
needlework	llll	
photography	ll	
woodwork	⊞ l	
yoga	⊞ llll	

3 Fill in the tally chart.

4 Draw a pictogram to show this data.

> **Remember the title and key.**

Holidays

The tables show how many students have visited the places listed.

Draw a pictogram to show each set of data. Use squared paper.

1 Holidays abroad

place	number of students
Jamaica	6
France	4
Portugal	3
Spain	8
America	5

Remember the title and key.

2 Holidays in the UK

place	number of students
England	7
Scotland	6
Wales	8
Ireland	2
Isle of Wight	3
Isle of Man	1

Remember the title and key.

Film watching

A survey asked people how much they spend on the cinema, videos and DVDs each week.

Here are the results:

12 people spend less than £5

8 people spend between £5–£10

16 people spend between £10–£15

18 people spend between £15–£20

6 people spend between £20–£25

2 people spend over £25

1 Put this information into the table below.

money spent on cinema, videos and DVDs	number of people

2 Now show the same information in a pictogram.

Use squared paper and remember to include a title.
A pictogram has symbols – give a key to show what they mean.

Represent information

Favourite takeaways

A survey listed favourite takeaways.

1 Record the data in the tally chart.

fish and chips	curry	chinese	pizza	curry	kebabs
pizza	burgers	curry	pizza	burgers	fish and chips
curry	fish and chips	burgers	pizza	burgers	kebabs
chinese	fish and chips	chinese	chinese	curry	curry
pizza	burgers	fish and chips	kebabs	curry	curry
fish and chips	burgers	curry	kebabs	fish and chips	pizza

takeaway	tally	total
fish and chips		
curry		
chinese		
pizza		
kebabs		
burgers		

2 Now show this data in a bar chart. Ask your tutor for some squared paper or graph paper.

> **A bar chart uses a scale to show the number of items. Remember to use a title and labels.**

Surveys

The tables below give some survey results.

Choose the best way to display the data.

Use squared paper.

1 Students attending classes

class	men	women
English	15	30
maths	25	15
computing	10	35
E-mail	30	30
ESOL	15	25
art	25	15
yoga	5	20

2 Favourite films of all time

film	number of people
The Great Escape	25
Titanic	15
Terminator	18
The Sound of Music	12
Saturday Night Fever	11
The Lord of the Rings	14

3 Preferred type of film

	thriller	romance	comedy	horror
men	25	5	55	20
women	15	35	45	15

Represent information

Now collect your own information.

Here are some ideas that you could use in college.

- What is the most popular colour?
- What is the most popular paper or magazine?
- Which is the most popular type of holiday: camping, hotel, B+B or self catering?
- What is the most popular make of car?
- How many hours do people spend in college?
- How many classes do people attend?
- How long does it take to get to college?
- What is the most popular type of potato: roast, baked, mashed or chips?
- What is the most popular fruit?
- What is the most popular vegetable?

Remember

Think carefully about how you ask people questions.

Write down any questions you will ask.

On a questionnaire, you may need to include 'other' boxes.

Ask a friend to try out your questionnaire.

Make charts or graphs to show your results.

Colour in your charts or graphs and label them.

Represent information skill check 1

p47-48, p50-51, p71

1 A group of friends wanted the following:

sandwich x 2	burger x 4	chips x 5	salad x 2
coke x 4	fruit juice	coffee	pop x 3
milk shake	wedges x 2		

One order was made for the food and one was made for the drinks.
Complete the lists:

food   **drink**

_____   _____   _____   _____

_____   _____   _____   _____

_____   _____

p80-81

2 Complete the pictogram for the food order.

Key △ = 1

_____

burger	△ △ △ △
chips	
salad	
sandwich	
wedges	

3 Show the drink order in a pictogram.

_____

Represent information

Represent information skill check 2

p72

1 Volunteer helpers have given their phone numbers to the college.
Order the list below so that tutors can easily find the phone numbers.

Max	472784
Debbie	754935
Brian	760430
Kath	764305
Francis	642950

volunteer   number

_____   _____

_____   _____

_____   _____

_____   _____

_____   _____

p74-75

2 Put the volunteer names on the timetable to show when they can help.
The times when they can help are shown below.

Max	Monday and Friday	1 pm–3 pm
Debbie	Monday, Wednesday and Friday	10 am–12 pm
Brian	Tuesday and Thursday	11 am–1 pm
Kath	Monday	1 pm–3 pm
Francis	every day	10 am–11 am

Volunteer timetable					
	10 am–11 am	11 am–12 pm	12 pm–1 pm	1 pm–2 pm	2 pm–3 pm
Monday					
Tuesday					
Wednesday					
Thursday					
Friday					

p83-85

3 Complete the bar chart to show how many days each volunteer helps.

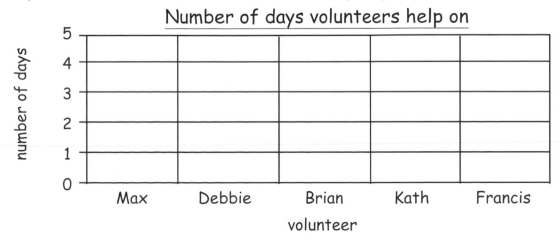

Number of days volunteers help on

Represent information   85

Represent information skill check 1 answers

1

food		drink	
sandwich x 2	burger x 4	coffee	pop x 3
chips x 5	salad x 2	fruit juice	coke x 4
wedges x 2		milk shake	

2 food order — Key △ = 1

burger	△ △ △ △
chips	△ △ △ △ △
salad	△ △
sandwich	△ △
wedges	△ △

3 drink order — Key △ = 1

coffee	△
pop	△ △ △
coke	△ △ △ △
fruit juice	△
milk shake	△

Represent information skill check 2 answers

1

Brian	760430
Debbie	754935
Francis	642950
Kath	764305
Max	472784

2

	10 am–11 am	11 am–12 pm	12 pm–1 pm	1 pm–2 pm	2 pm–3 pm
Monday	Debbie, Francis	Debbie		Max, Kath	Max, Kath
Tuesday	Francis	Brian	Brian		
Wednesday	Debbie, Francis	Debbie			
Thursday	Francis	Brian	Brian		
Friday	Debbie, Francis	Debbie		Max	Max

3

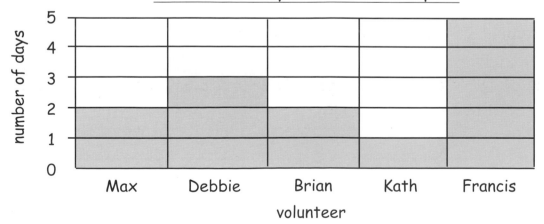

Number of days volunteers help on

1

Shop opening times	
Monday	8:30 – 7:00
Tuesday	9:00 – 7:00
Wednesday	9:00 – 12:00
Thursday	9:00 – 7:00
Friday	8:30 – 7:00
Saturday	8:30 – 5:00

When does the shop open on Wednesday? _____

The shop has a delivery.

baked beans x 12	lettuce (box)
bread x 10	milk x 24
cheese x 3	soup x 6
curry (tins) x 2	tea (packs) x 5
grapes (bags) x 6	yogurt (packs) x 8

2 Circle the word that best describes how the list is ordered.

      numerically   alphabetically   by date

3 How many packs of yogurt are there? _____

4 How many boxes of lettuce are there? _____

5 There are 12 of one item. Which is this? _____

6 Some of the items need to go in the fridge.
List the items for the fridge.

Mack test

The shop keeper makes a list of all the items in the fridge with their sell-by dates.

cream 12th Sep	yogurt 21st Sep	yogurt 18th Sep
milk 17th Sep	ham 16th Sep	paté 19th Sep
cheese 11th Sep	milk 13th Sep	chicken 20th Sep

7 Rewrite the list in date order.

_____   _____

_____   _____

_____   _____

_____   _____

_____   _____

_____   _____

_____   _____

_____   _____

8 It is 14th September. Circle all the items that are past their sell-by date.

A delivery of ready meals comes to the shop.

vegetable lasagne x 5   chicken curry x 3   vegetable curry x 2

roast lamb x 2   bacon quiche x 1   bean casserole x 4

Mock test

9 Complete the pictogram of the ready meal delivery.

<u>Ready meals</u>

Key ☐ = 1 meal

roast lamb	☐ ☐
bacon quiche	
vegetable curry	
chicken curry	
bean casserole	
vegetable lasagne	

10 Some of the meals are vegetarian and some are not.
Make a list of the meals that are vegetarian.

Vegetarian meals

_____

_____

_____

11 Below is a plan of the shop.

Tins are to the right of the bread. Label the counter **T**.
Vegetables are opposite the door. Label the counter **V**.

The fridge for ready meals is between the vegetables and the fridge 2. Label it **F3**.

The milk fridge is on the left of the door. Label it **F1**.

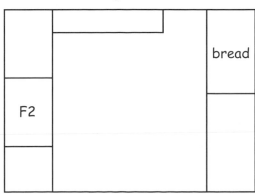

door

Mock test

1 It is planned to build a new car park at college.
   A questionnaire is needed.
   Tick the question which is most suitable.

   When do you come to college?
   Would you use the college car park?
   Do you have a car?
   Can you use the bus or train to get to college?

A student checks people parking cars between 9 and 10 am.
Here are the results.

visitor	tutor	student	student	visitor
tutor	tutor	student	tutor	student
student	student	student	student	tutor

2 Complete the tally chart.

	tally	total
student		
tutor		
visitor		

3 Make a pictogram to show the results. Give a title.

_____   Key ◯ = 2

student	
tutor	
visitor	

4 Here is a plan of the new car park.

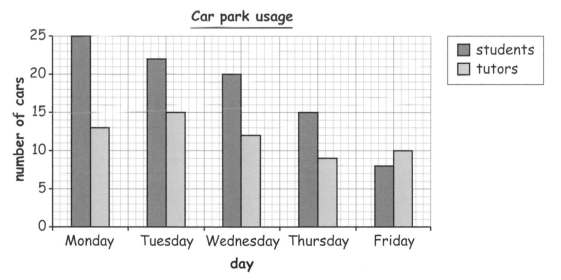

Key
student park = S
tutor park = T
visitor park = V
pay machine = P

The pay machine is to the left of the entrance.
Show this on the plan with a square. Label it P.
The visitor parks are to the right of the entrance.
Label these as shown on the key.
The tutor parks are opposite the visitor parks.
Label these as shown on the key.

Here is a chart.

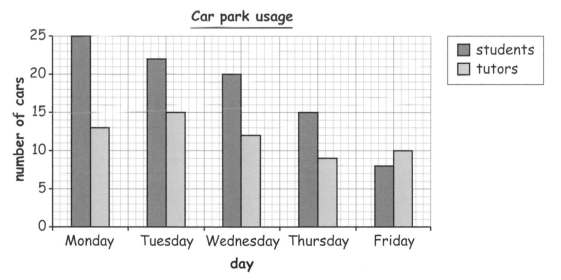

5 On which day did more tutors park than students? _____

6 Which day has the least number of tutors parked? _____

7 How many students parked on Wednesday? _____

8 How many students and tutors used the car park on Tuesday? _____

Mock test

9 What is missing from this chart? _____

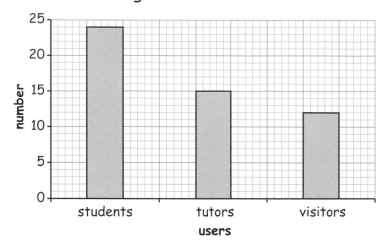

The table gives the car park users during one week.

	student	tutor	visitor	total
Monday	24	14	3	
Tuesday	32	17	2	
Wednesday	25	15	6	
Thursday	22	12	3	
Friday	6	12	7	

10 Fill in the totals column.

11 How many students used the car park on Monday? _____

12 On which days did 12 tutors use the car park? _____

13 When did more visitors use the car park than students? _____

14 These tutors will use the car park on a Saturday open day.
Use the key to find at what time they will park.

Madge	**M**	Bill	**A**	John	**E**	Jane	**M**
Jill	**E**	Jo	**E**	Jaj	**M**	Cath	**A**
Ben	**M**	Paul	**M**	Phil	**A**	Jess	**E**

Key
morning = **M**
afternoon = **A**
evening = **E**

Complete the chart.

Handling data mock tests

Extract information

Page 1

Find the phone number
1. 424 192
2. 425 839
3. 415 371
4. alphabetically
5. 243
6. room 11
7. room 14
8. numerically

Page 2

Library
1. 9:00 am
2. Friday
3. Sunday
4. Friday
5. 3:00 pm
6. Monday and Tuesday
7. 10:00 am
8. 4:30 pm

Page 3

Shopping list
1. 4 cans of lager
2. loaf of bread
3. 3
4. 1
5. 5
6. 3
7. randomly

Page 4

Hospital stay
1. randomly
2. underwear x 3
3. 3
4. pyjamas
5. conditioner
6. 3
7. toothbrush
8. 2

Page 5

College notice board
1. L8
2. B2
3. D.I.Y.
4. L3
5. alphabetically
6. there is no cookery

Page 6

Ordering lists
1.
Brian	483 4902
Debbie	753 5829
Fiaz	571 4525
Imra	859 4623
Kath	728 6472
Max	839 7591
Rhonda	573 3832

2.
1st Jan	Meet Bev
2nd Jan	Cinema
3rd Jan	Shopping
5th Jan	Term starts
10th Jan	Dad's birthday
12th Jan	Ice skating

3. Aled Carpets
   Cover Up
   Eversure Carpets
   First Floors
   Imagine Floors
   Style and Tile

Page 7

New authors
1. Delia Tyre
2. alphabetically by surname
3. Over the Hill
4. Amos Murray
5. Forest Games
6. Jen Moseley
7. True Dawn
8. In the Deep
9. Reggie Charles

Page 8

Netball league
1. Castle Vale
2. Hilltop A and Rovers A
3. Castle Vale
4. no
5. Rovers B
6. 5
7. West Lea
8. Beechy Head

Page 9

Dog training
1. Jess
2. Casper
3. George T
4. Taylor
5. Eliza
6. Taylor
7. 22/9
8. 22/9
9. alphabetically by owner

Page 10

Holiday in New Zealand
1. 17 °C
2. 15 °C
3. Queenstown
4. Auckland, Tauranga, Milford Sound, Christchurch
5. March
6. Auckland
7. Milford Sound
8. no

Page 11

Pizza Place
1. £3.50
2. £5.30
3. £7.80
4. Seafood
5. 8 inch Pepperoni Plus
6. they cost the same
7. Seafood/Vegetarian

Page 12

Bus times
1. 9:00 am
2. 10:00 am
3. 6:50 pm
4. 7:00 pm
5. no
6. yes
7. 1st bus
8. 1st bus
9. 3rd bus

Page 13

Banking
1. by date
2. £7.50
3. debit
4. £131.77
5. 8th Feb
6. 9th Feb
7. 9th Feb
8. cash withdrawal

Page 14

Mileage chart
1. 88 miles
2. 215 miles
3. 206 miles
4. Manchester and Sheffield
5. 172 miles
6. Bristol and Oxford
7. York
8. Liverpool and Sheffield
9. Sheffield

Page 15

More distances
1. 62 miles
2. Perth
3. 43 miles
4. 67 miles
5. Fort William and Glasgow
6. Aberdeen
7. Inverness
8. Perth
9. Glasgow

Page 16	Page 17
Ice rink	**Teaching room**
1 2	1 7
2 seating	2 3
3 café	3 maths shelf
4 Checker's bar	4 computer 1 and
5 Top Skate shop	cupboard
6 office	5 printer 1
7 exit	6 computer shelf
8 Top Skate shop	7 table

Page 19

Leisure centre questions

1 2	2 8
3 bowling alley booking	4 13
5 screen 1	6 2

7 Go ahead from the main entrance, turn left past the bowling alley booking, then right past some of the lanes. Turn right past the toilets. Go straight on to screen 2.

8 Go ahead from the main entrance, turn left past the bowling alley booking to the shoe hire.

Page 21

College plan questions

1 reception
2 staff room and English 1
3 technical support
4 interpreters
5 crèche and German 2
6 Hindi
7 French 1 and Japanese

Page 23

Castle plan questions

1 gatehouse	2 4
3 well house	4 3
5 the keep	6 the wall walk
7 museum	8 Saxon wall
9 shop	10 postern gateway

Page 25

Finding your way

1 Bournville Lane
2 Sycamore Road and Maple Road
3 no
4 yes
5 Laburnum Road
6 Acacia Road
7 Maple Road and Acacia Road
8 This is one possible route: Turn left on to Bournville Lane. Turn right onto Linden Road. Turn right onto Raddleburn Road. The hospital is on the left.

Page 26

Teaching room measurements

1 6 m	2 7 m	3 3 m
4 13 m	5 no	6 1 m
7 3 m		

Page 27

Car colours

1 4	2 6	3 2
4 red	5 white	6 green

Page 28

Drinks

1 Key ⊍ = 1 cup

   Title should include: **Drinks sold on Monday**

tea	⊍ ⊍ ⊍ ⊍ ⊍
coffee	⊍ ⊍ ⊍ ⊍ ⊍ ⊍
chocolate	⊍ ⊍ ⊍
soup	⊍ ⊍ ⊍ ⊍
soft drink	⊍ ⊍ ⊍ ⊍ ⊍ ⊍ ⊍ ⊍

2 soft drink	3 4
4 3	5 8
6 chocolate	7 tea

Page 29

TV viewing

1 Key picture = 1 person. (e.g. 🚹 = 1 person)

   Title should include: **Soaps people like best**

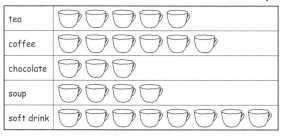

2 8	3 2	4 9
5 Neighbours	6 Emmerdale	7 29

Page 30

Television sales

1 20 inch screen	2 mini
3 5	4 25
5 20	6 10
7 35	8 15

Answers

Page 31

Holiday choices
1 2 people
2 Brighton
3 Scarborough
4 10
5 20
6 16
7 Bournmouth and Fishguard
8 70

Page 32

Cars
1 Cars in the car park
2 Volkswagen
3 Vauxhall
4 5s
5 1 car
6

make of car	number of cars
Peugeot	14
Volkswagen	18
Vauxhall	9
Ford	15
Nissan	12
total	68

Page 33

Bills
1 cost of electricity and gas
2 £s
3 £10s
4 March and April
5 June
6 January
7 £6
8 £40
9 £10

Page 34

Café drinks
1 Drink sales
2 2 drinks
3 fruit juice
4 soup
5 and 6

type of drink	number of drinks sold
fruit juice	36
coffee	35
tea	25
soft drinks	24
chocolate	16
soup	12

Page 35

College courses
1 college courses
2 4
3 5
4 English
5 it has the largest bar
6 7
7 2
8 27

Page 36

Smoking survey
1

age of smokers	number of people
20 and under	19
21–30	34
31–40	35
41–50	25
51–60	20
61–70	18
over 70	15

2 53
3 33
4 45
5 31–40
6 166

Page 37

Fruit
1 people's favourite fruit
2 banana
3 pear
4 10s
5 2s
6 1
7 70

Page 38

Films
1 the number of films sold
2 thriller
3 10s
4 comedy
5 32
6 children's
7 more adults buy films

Page 39

Smokers
1 21–40
2 under 21
3 41–60
4 under 21
5

smokers	number of men	number of women	total
under 21	16	22	38
21–40	26	26	52
41–60	30	24	54
over 60	27	16	43
total	99	88	187

Page 40

Classes
1 classes attended by students
2 one shows men and the other shows women
3 women
4 English
5 maths
6 no
7 90
8 72
9 ESOL

Page 41

What is missing?
1 title
2 is right
3 axis title: **means of travel**
4 scale

Sort and classify

Food and drink
food = bread,
    carrots,
    crisps, fish,
    curry, pizza,
    spaghetti
drinks = lager, coke,
    coffee,
    milk, water,
    lemonade,
    tea

Page 48

Recycling
Oxfam = shirts,
    t-shirt,
    shoes, hat,
    sandals,
    jumper,
    shorts,
    trousers
Paper = post card,
    letter,
    magazine,
    books,
    this book,
    calendar

Page 49

Food shopping
baker = bread, cake,
    doughnuts,
    sausage roll
butcher = bacon,
    chicken,
    sausages
greengrocer =
    bananas, apples,
    lettuce

Page 50

Tools
garden shed =
mower, rake,
wheelbarrow,
garden fork, hose
garage = hammer,
screwdriver, drill,
saw, ladder

Page 51

Shapes
square, rhombus,
circle, rectangle,
cross should be
coloured in

Page 52

Vehicles
1 coloured red =
    motorbike, penny
    farthing, bicycle
2 coloured blue =
    tricycle,
    3-wheeled car,
    motorbike with
    sidecar
3 4-wheel vehicles
    are not coloured

Page 53

Sort them
1 subject
2 contents
3 colour

Page 54

On the hour
clocks circled = 3, 5,
6, 11, 13

Page 55

Jumbled shapes
blue = 2, 12
red = 3, 5, 6, 7, 8,
    11, 13
green = 1, 14

Page 56

Meals
1 5   2 7   3 4

Page 57

Clothes
1 3
2 no
3 blue
4 green and black
5 blue
6 1
7 1
8 yes
9 yes

Page 58

Dog classes
1 16   2 10   3 3   4 3   5 agility
6 6   7 62   8 agility and heel work

Page 59

Favourite drinks
Ask your maths tutor to check this for you.

Page 60

Fast food
Ask your maths tutor to check this for you.

Page 61

Holiday park

holiday park	tally	total
Coast	卌 ‖	7
Bay	卌 卌 ‖‖	14
Hill top	卌 卌 ‖	12
Town	卌 卌 ‖	11
other	卌 ‖‖	8

Page 62

Music

music	tally	total
classical	卌 ‖	7
rock	卌 卌 ‖	11
jazz	卌 卌 ‖	12
pop	卌 卌 ‖‖	14
other	卌 ‖‖	8

Page 63

Questionnaire
improvements need to include:
have clear age groups
more choice for how often you borrow from
the library
choice for time of day
yes/no boxes for do you borrow books, do you
borrow CDs and do you use the computer
space to write the book type you prefer or a
choice of types

Page 64

Pets
1 What pets do you have?
2 verbally
3 all or any of these
4 during the day
5 at a supermarket, on the train and at work

Answers

Represent information

Writing a list
apple x 6
doughnut x 2
lettuce
milk x 3
beer x 5
banana x 4

More lists
1
Bill 642814
David 079563728463
Leila 672453
Mohammed 346282
Paul 673828
Rajaa 079774435534
Soumia 382848
Vicky 678929

2 p4, p7, p8, p11,
p13, p16, p18, p21

3 1 March
   5 March
   12 March
   21 March
   29 March
   2 April
   9 April
   17 April
   24 April
   26 April

Dog kennel jobs
1 Sunday
2 3
3 3
4 14
5 shaded section for
  each person: walk
  dogs x 7, clean
  runs x 3, see vet
  x 1, clean visitor
  room x 1, feed
  dogs x 3

  One of each section
  marked **see vet**,
  clean visitor room
  and **feed dogs** are
  not shaded because
  they cannot be split
  completely fairly.

Office spending

heading here e.g. item	week 1	week 2	total
envelopes	£2.40	£1.50	£3.90
stamps 1st class	£3.00	£1.50	£4.50
stamps 2nd class	£2.10	£2.10	£4.20
paper	£2.40	£3.00	£5.40
pens	£3.00	£2.00	£5.00
pencils	£1.50		£1.50
milk	£1.42	£1.42	£2.84
coffee		£5.95	£5.95
sugar	£0.68	£0.68	£1.36

Timetable

	Monday	Tuesday	Wednesday	Thursday	Friday
9–10	Spot		Lady		
10–11	Spot				Ben
11–12			Sandy		Ben
12–1			Sandy		
1–2	Bruno		Bob		Trixie
2–3	Prince				Trixie

Table top sale

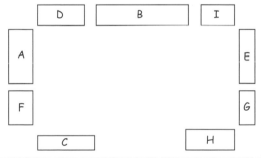

Design a kitchen
These shapes on the grid with labels or a key:

Library
blue = Basic Maths Number, Numbers Made Easy, Fun With Numbers,
       Shapes and Numbers, Number Crunching
orange = Basic Maths Measures, Measuring is Fun, Measuring Tools,
       Measure Yourself, Ways of Measuring
green = Basic Maths Handling Data, Charts and Graphs, Find the Data,
       Check that Chart, Handle the Information

Answers

Page 77

Drawing plans

Ask your maths tutor to check this for you.

Page 78

Class charts

1.

classes	tally	total
maths	卌 |	6
English	卌 |||	8
computing	卌 ||	7
ESOL	||	2
typing	卌	5

2. Title (e.g. number of classes)

Key △ = 1 class

maths	△△△△△△
English	△△△△△△△
computing	△△△△△△
ESOL	△△
typing	△△△△△

3.

leisure classes	tally	total
art	卌 ||	7
needlework	||||	4
photography	||	2
woodwork	卌 |	6
yoga	卌 ||||	9

4. Title (e.g. number of leisure classes)

Key △ = 1 class

art	△△△△△△△
needlework	△△△△
photography	△△
woodwork	△△△△△△
yoga	△△△△△△△△△

Page 79

Holidays

1. Title (e.g. students holidaying abroad)

Key △ = 1 student

Jamaica	△△△△△
France	△△△
Portugal	△△
Spain	△△△△△△
America	△△△△

2. Title (e.g. number of students holidaying in UK)

Key △ = 1 student

England	△△△△△△
Scotland	△△△△△
Wales	△△△△△△△
Ireland	△△
Isle of Wight	△△△
Isle of Man	△

Page 80

Film watching

1.

money spent on cinema, videos and DVDs	number of people
less than £5	12
£5–£10	8
£10–£15	16
£15–£20	18
£20–£25	6
over £25	2

2. Ask your maths tutor to check this.

Page 81

Favourite takeaways

1

takeaway	tally	total
fish and chips	JHH II	7
curry	JHH IIII	9
chinese	IIII	4
pizza	JHH I	6
kebab	IIII	4
burger	JHH I	6

2

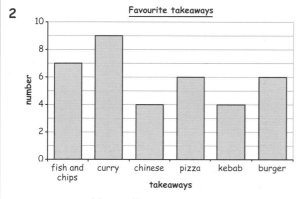

Page 82

Surveys

1

2

3

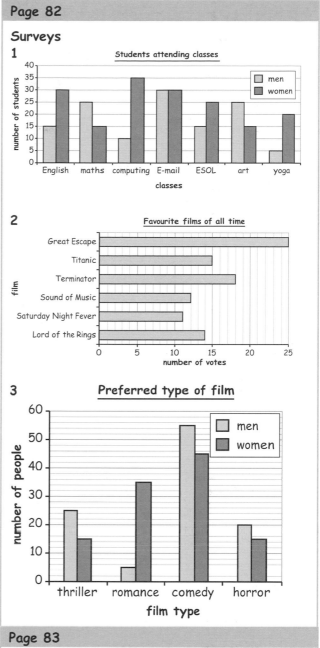

Page 83

Collect, record and display
Ask your maths tutor to check your work.

Handling data mock tests

Handling data mock test 1
1 9:00 am
2 alphabetically
3 8
4 1
5 baked beans
6 cheese, milk, yogurt
7 11th Sep cheese
  12th Sep cream
  13th Sep milk
  16th Sep ham
  17th Sep milk
  18th Sep yogurt
  19th Sep paté
  20th Sep chicken
  21st Sep yogurt
8 cheese, cream, milk from 13th Sep
9

roast lamb	□ □
bacon quiche	□
vegetable curry	□ □
chicken curry	□ □ □
bean casserole	□ □ □ □
vegetable lasagne	□ □ □ □ □

10 vegetable curry, bean casserole, vegetable lasagne
11

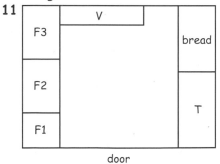

Handling data mock test 2
1 Would you use the college car park?
2

	tally	total
student	⊞ |||	8
tutor	⊞	5
visitor	||	2

3 People parking cars between 9 and 10 am

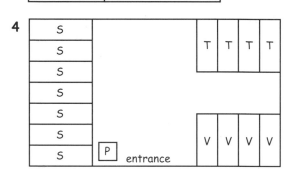

4

S					T	T	T	T
S								
S								
S								
S								
S					V	V	V	V
S	P	entrance						

5 Friday
6 Thursday
7 20
8 37
9 the title is missing
10 Monday 41, Tuesday 51, Wednesday 46, Thursday 37, Friday 25
11 24
12 Thursday and Friday
13 Friday
14 One solution is:

Car park times

number parking	morning	afternoon	evening
5	Madge		
4	Ben		Jill
3	Paul	Bill	Jo
2	Jaj	Phil	John
1	Jane	Cath	Jess

times of day

Answers